# Holdfast

Annette Skade

CHANNEL

Holdfast

Published in Dublin, Ireland by *Channel*

Copyright © Annette Skade, 2024

The right of Annette Skade to be identified as the author of this work has been asserted in accordance with the provisions of the Copyright and Related Rights Act, 2000.

All rights reserved. The material in this publication is protected by copyright law. Except as may be permitted by law, no part of the material may be reproduced (including by storage in a retrieval system) or transmitted in any form or by any means, adapted, rented or lent without the written permission of the copyright owners.

ISBN 978-1-0686370-1-8

Printed by Print Media Services, printmediaservices.ie

Design and layout by Cassia Gaden Gilmartin

Cover art and illustration from *seaweed* by Marina Dmitrik, metrolstudio.com

Connect with us: www.channelmag.org | info@channelmag.org
facebook.com/ChannelLiteraryMagazine/ | x: @Channel_LitMag | instagram: @channel_mag

*Channel* receives financial assistance from the Arts Council.

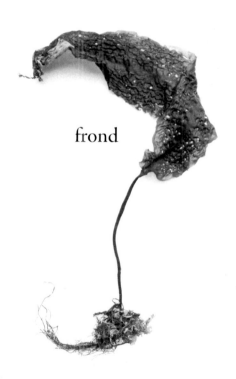
frond

"I live an extremely retired life, next to complete seclusion & in such a situation it is rather doing me a kindness to excite exertion."

fronds spread to take in sunlight
a dark mass under the surface
pebbles roll from under my feet
cold encases
this is my first feel –
gelatinous
smooth
almost animal
seaweed

sea-weed: a word linking land and water
what grows and the force it thrives in
what do you see when you hear this word?
one small spore floats
in the tangle of your thoughts
let your thoughts grow around this
you are on the seashore

Ellen's skirt billows and spreads
surface glare pulls hands to eyes
tugs her beyond the pattern
of scintillating pockmarks
towards the horizon
her eyes flick back to tide line
hanks of hair stream loose from their pins

how her world shrinks

We are all called home, tethered to a radius
within walking distance. Checkpoints
block our single road east and west. Hills behind us,
sea in front. We are not so much curbed by this
as by a prayer to take care; to have care.
The letter calling Ellen back has similar weight.
The screws turn. She is pressed within the paper.

I change walking routes to ones more solitary,
down to rocky inlets, seaweed piling into crevices
or thrown onto sheep tracks. I go up onto the hill's
slippery bogs. I am stooping to examine a tiny flower,
dried and dead, still held on a paunch of moss,
when I first feel Ellen. I straighten to stare
down on a landscape curving round the bay

like the crook of an arm, remember her home
rested in that tender part inside the elbow,
hidden from the worst of the weather.
Yet the scalpel slid into this soft place,
opened the veins of mother and brother,
drained each to a shadow
on advice from experts wholly in the dark.

Ellen looks seaward at a bay coated gold,
depth of sky measured in the criss-cross
of birds' flight. She sees what I see: algae
in dark furrows like a rich ploughed field,
ready to bounce back with the incoming tide.
Mainly egg wrack: fronds in long strings,
air pockets to keep the plant buoyant.

Her eyes sift the weighted tide
for unusual shades or frond shapes.
Some she takes back
to tease out details which identify the plant,
to examine the position of capsule or "seed"
under the microscope, laying aside rarities
for her next letter.

Their correspondence speaks of science
and poetry, words cast wide across the water.
They swap opinions on species,

on Dante and the newest Byron.
Metaphor and Botany: forays
in the overlay of sameness and difference.

She enquires how loved ones fare,
how much they draw nearer to life or to death.
His reply brings diversion "at a time when domestic troubles
lean so heavily upon me."
Her mind floats on pockets of enquiry:
the death of a friend and the drawing of *confervae* –
"new ones I hope" –
in the same paragraph.

Inside a triangulation of seashore, hill and wood,
I locate Ellen. She documents her passage
in place names and Latin:
*Atra* in rivers at Glengarriff,
*Fucus capillaris* at Ardnaturris,
*Fucus amphibius* in Ballylickey Cove,
*Laceratus lingulatus* near Eagle Point.

guides to identification
divide seaweed into three colours:
brown / green / red /
they share habitat
the same pages in the natural history book
but their lineage is not the same
green and red are related to mosses
ferns, flowering plants; brown
akin to certain types of fungi
algae break free of taxonomy

Ellen shifts boundaries
notes the similarity of sea plant and furze
in terms of abundance and colour

adjectives run into each other
pinky red purplish red
glossy brown reddish brown
dark blue green colour
olive brown beautiful green

"Its colour when recent is much more brilliant"
colours change –
in spring
from age
after drying

a postscript
"2
          pink coloured
  parasitic      species
             little
beauties"

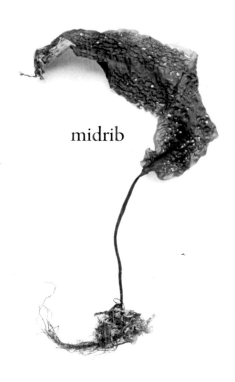

midrib

"Your *rivularia vermiculata* I am inclined to think not the same as my plant."

In season
a thread can be followed
through the centre of the frond.
Sometimes faint,
sometimes swollen to a ridge.
Midrib: the central "vein" of the lamina,
neither vein nor bone,
present only in certain types.
*Fucus laceratus*,
"a large & beautiful veined variety"
of which she sends a single plant
since she meets it rarely.
Another, round and full when fresh,
flattens within minutes to show the inner thread.

The midrib of tongue weed when young
is barely discernible,
raises with age like blue
on the back of a papery hand.
In winter the leaf falls away,
leaving a single spike.

The curling strokes of Ellen's hand
in ink browned and thinned with age
are more than the sum of kink and curl.
Her hand bleeds narratives of death
and lengthy sufferings, medicines
that soothe but cannot cure.
Now fetch her letter to your lips.
It tastes of salt.

Samples – two centuries old – pressed
on acid-free herbarium paper,
stacked in folders inside indexed lockers.

I am instructed on how to lift each page:
corner to diagonal corner between finger
and thumb – less likely to bend the sheet,
crack the specimen.
Entire small reds float in pools of colour
they themselves have secreted. They glide
towards me: jellyfish on parchment.

Within stacks of samples
I come across ghosts.
The brown stain upstrokes of stipes,
imprint of curving fronds, witness to
the presence of parts of the thallus
otherwise gone. Specimens glue themselves,
exude so much mucus that they stick fast,
rippling the paper.

The bladder wrack has turned to graphite,
bladders flat as wax seals.
Serrated wrack shows powdery black,
a pale line chalked along the midrib
meandering from holdfast to tip.
In days under weighted card,
years glued to acid free paper,
within stacks of files,
all along its crooked spine,
the plant has sweated centuries of salt.

The Atlantic coast is battered by waves
and scoured by salt. Salt rimes cars,
rusts metal railings. It coats our skin and lips.
Come May, the sprinkle of summer on your tongue
is straightaway rinsed by a salt-laden gale.
The lift in your heart is premature.

Wind reddens skin, turns new growth black,
baffles any train of thought. Ellen longs for
the early light of summer mornings – more time
to collect or write letters before breakfast.
Fine days, she takes a boat to rocks and strands,
returns, samples cosseted, hair thick with salt.

She makes "parcells", lays the fragile plant
on glass slides – easier to lift for the microscope.
Veined algae branch across the surface,
drawn to the memory of sand,
the slow flow of the glass.
Lamina spread like fingers
on the gloss of a flat calm sea.
Seaweed shares names
with parts of the body: hand
 – hand leaf bearer –
liver, ear, gut, kidneys, eyelash.
Seaweed straddles sea and land,
animal and vegetable.

Folklore tells of rock pool creatures,
half seaweed, half fish. Look close
at bladder wrack in Spring: wavy blade
echoes swimming action, yellow ridge
along the length blunts to swollen lips,
yellow eggs like bulbous eyes
on each side of the midrib.

Saltwater transforms. Each time you immerse
you are changed. Waves pummel
and loosen your spine, refraction
dislocates your limbs. You spread
wide on the push below you.

The sea is cool and delicious.
I brush over a bank of dillisk,
swim past the queasy sensation of grab
into sea silky with invisible spores.

Back on the upper shore I dry off,
raise arms above my head to soak up sun,
brush the tips of a saucer-sized clump:
fork weed, brown as the rock it sits on.
The tops break off in my fingers,
branch-like fronds the size of my nail
grown lifeless as a herbarium sample,
fallen prey to sun and salt.

Ellen's letters tell of headaches, bilious attacks.
Inside, the walls of her organs flare.
They send for the physician.

She sickens and rallies. Her spirit rises from the bed,
beats against the room like a swallow trapped,
breaking wings against the window.

She can only creep behind, holds on to an arm,
a hand, the furniture. She drops into a chair,
looks onto rain so thick everything is wrapped in it.

Open the window a crack. Hear the stream in spate.
Sound works like the heel of the hand on breastbone,
pumping: breathe, breathe, breathe.

"I am now so much better, tho' I have not been out for
more than two months untill this day." Around here,
all rivers find their way to the sea. So does she.

stipe

"I have spent many happy hours creeping among its rocks & never quitted it without regret."

Seaweed belongs to the edge.
She says she is happiest outdoors,
spends what time she has hunting
species on tops and edges.
Her favourite plant, *Rivularia*, is only to be got
where streams open out into sea.

Algae fills littoral, sub-littoral space.
It finds ways to hide from the hack of the waves
in rock pools above the low spring tide line
or submerged just below. Some can only thrive
in a strip of shore not much wider than two metres,
survival as precarious as our own.

I am coughing, chest tight. But the test
is negative. This time carrageen will shift it.
I take a handful of scrunched-up plant,
steep, boil until the liquid thickens, strain
and pour. I add honey and lemon
then drink it down.

I feel the viscous liquid coat raw throat,
recall how seaweed fights harm with silk,
exudes a coating to turn the waves' blunt force.
This cure was foraged from shore,
handed on by word of mouth
long before Ellen was born.

Before I have the names
I make metaphors on the shoreline:
this one is combat-gear in ribbons,
this a tiny stook of stubby brass twigs.
Thong weed fans out like the hair

of a woman drowned,
kelp stipes are silted bones.

This rose of metallic red gift ribbon
ravelled by the scissors' blade
I later learn is tongue weed.
Here is bladder wrack,
egg wrack.
Serrated wrack is deep bronze,
orange tips fat with spores,
ready to burst on the swell.

The names I learn are common names,
not containers for categories
like Latin nomenclature.
They are metaphors reflecting how algae
slip through fingers.

Ellen's answer is to paint. A ribbon of red
submerges, mingles to pink.
She removes brush from jar,
holds it poised above a palette
dotted with white / red / a little brown,
shades mixed to create
exactness.

Her gaze is forensic.
Of *Rivularia vermiculata* she says
"I never saw anything more beautiful than
its magnified appearance."
She paints what is:
all that the sample has in common with others.
Then, with the tip of the narrowest camel hair brush,
lifts the parts that make the plant distinct
into the hyper-real.

Each detail takes on the look of something just below
the surface of clear water –
a fraction more itself.

To lift the contents of the archive box
is to touch what Ellen has touched.
I peer into the past through a hole
shaped from Ellen's possessions.

A letter to her brother is cross-hatched.
Lines bottom to top / left to right
on one precious piece of paper.
The criss-crossed words are hard to read,
point to all that is unsaid,
so much of a life still left unknown.

Like a secret, the letter folds in on itself,
franked 4 o'clock No.19 1808EV,
wax seal cracked open.
On the reverse:
*BANTRY 16*
*Emmanuel Hutchins*
*Kings Ington*
~~*London*~~

The box contains three poetry books.

*Childe Harold*, much used
by Ellen and others, has lost its cover.
Time or damp has puckered the fore-edge.
The page undulates like a Venetian canal.

Tasso's *L'Aminta di Torquato*
(dated 1812) bears Ellen's level signature.
A tear in the buff paper cover reveals a spine

shored up with scraps from a dictionary.
English-Italian.

*To cut and mangle – mutilo*
*To cut afunder – refcindo*
*To cut or lop trees*
*To cut one's hair*

Dante's *Comedia* fits into the hand,
tooled leather in deep red and cream.
Ellen speaks of her delight in this poetry,
and the horror of it.
I ease the book open: an action
as intimate as cutting someone's hair.

I slide open the drawer of the pipped-oak archive cabinet.
Under polished glass, Ellen's painting of *Fucus asparagoides*
recalls the most delicate red fern. A stipe a millimetre wide
branches and re-branches to the slimmest of fractals.

Deeper red on the stipe's right edge pales to the left.
This change of hue embeds the plant and raises it
to the eye. She has selected one thumb-length frond
to magnify at the bottom of the page.

She enlarges to aid research, but lets the beauty in,
leaves bare watercolour paper to shine through
in thousands of minute crescents, creating
dapples across the surface of the thallus.
.
Her brush exposes details invisible to the naked eye:
curved spikes along a central rib. At the base of each,
a spatula-shaped stub holds clutches of spores.
In life, each capsule is smaller than a pinhead.

Algae is open to whatever motion the sea brings,
moulds itself with a plasticity
beyond that of most land plants.
It changes shape to fit where it grows.
The broad, fleshy leaves of one sample
signal a life away from wave action,
another wiry stalk of the same species
is stripped to withstand pounding rhythm,
hard to identify but for two fleshy ears
pricked at each tip.

Built to yield to the force of the ocean,
it meets deluge with slippery silk,
sieves salt water through miniscule fronds
or fathoms of kelp forest.

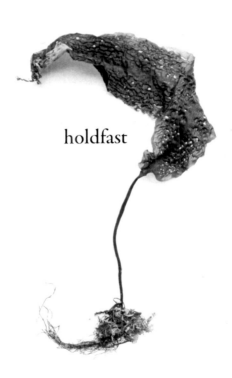

holdfast

"Send me a moss anything just to look at."

The old people say never
turn your back on the sea.
You think the tide is ebbing
and are caught in the flood,
no way to keep your footing.
You are swept into the sea's mouth.

Harvesting by hand still happens.
A day of hail and gusts, broad
rain-wiped sky. A man out on the spit
hooks heaps of kelp, sea racing
on three sides. This weed
is destined for a factory in the North,
becomes the silk in tights and stockings.

They used to make rafts,
roped the weed to float back to the village,
or rowed for an hour to get to the rock
where the strapweed is slender,
heavy with sap. Seaweed to sell
or to put on the fields.

Algae do not root. A holdfast
in the form of disc, branch or claw
anchors to rock.
It weathers the power of wind,
the fetch of a wave building
from the coast of America,
by cleaving to the surface
of something as small as a pebble.

Ellen brings a hammer to take samples.
Holdfasts grip so tight that to wrench them away
can destroy the thallus. A herbarium sample
of serrated wrack, taken a century ago,

still clings to a chip of stone.
It is speckled with uniform white spots.
The lens reveals each as an entire pearly shell.
This plant on a tray-sized page is a raft
for a colony of miniscule snails.

The date and place are recorded.
My mind travels to that same spot,
hopes to find the same plant,
raises a hand lens to search for flecks of white,
miraculous living spirals
prey for so long to discharge and run-off.

Spring tides.
The road eastward mimics the coastline it follows:
storm-driven waves salt the asphalt with shells,
small stones. Dark patches of seaweed are hurled
over flood defences.
In the town where Ellen is buried
thresholds are sandbagged,
wrack drapes the harbour wall near the square.

In memory of the last request
she would ever write to him,
I bring moss to her burial place.

Perennially twenty-nine, she has lain here
for more than two hundred years.
There is a tang of salt and stillness
but no headstone, no exact location.
A blue plaque records name, date,
her service to cryptogamic botany,
the fact that she is buried somewhere here.
The words shrink a life to fit a small surface.

I could go down to a nearby strand, pick
a dry piece of seaweed and wait for it
to quicken when the air moistens –
a sign of oncoming rain. For years I've kept
a plant as weathervane, hung it from a nail.
Is there rebirth for Ellen?
She lies in roots and rhizomes, rests
in the web of this delicate ecosystem.

Seaweed attaches to artificial structures
as well as to rocks: boat moorings and piers,
the pontoon where the ferry berths.
In time the waterline of this town will rise,
foreshore pushed up and back.
Algae will cover concrete and brick,
the graves I stand among becoming upper tidal.
Bones will mix with kelp stipes and thong weed.

He learns of Ellen's death by letter.
Her sister-in-law writes
"My beloved Miss Hutchins breathed her last
in my arms the ninth of this month."
She speaks of practicalities: Ellen's desire
that a parcel of plants be sent to him.
She confesses herself unequal in her grief
to tell more of "our sweet friend."

Those of us who know loss
see presence and absence,
touch and the lack of it
interplay between these lines.
We blink our way out
like people too long in the dark,
our movements slower, more careful.
At least for now.

Each piece of seaweed urges you
to hold fast, to take one small place
for your own, one scrap of ground,
one book, one unusual sample,
one meticulous illustration,
one hand in your hand.
It bids you trust to the swell,
to the rise and fall of your own breath
saying *here*,
                saying *home*,
          saying *hold on*.

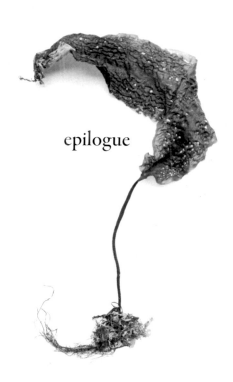

epilogue

"The Bay has been all summer full of Whales, they came in after Herrings & sprats."

Come back with me full circle
to the hilltop where we first felt Ellen breathe.
The sea blends into sky, light
draws the eye to the stretch of water.
From this vantage point
there are chalk marks on the sea's grey,
powdery graffiti undulates across the straits.

Take in the view. See the scatter
of fishing boats, ferries to the islands,
small yachts and sailboats.
You can just make out
the far tanks of the oil terminal
gapped like ill-fitting kerbstones,
the tanker towing horizon towards it.
We are too high to hear the rumble of engines,
too far away to smell the diesel,
to take much notice of machine harvesters
forging a path towards the bay.

Ellen is alive to the tiniest epiphyte
growing from the stipe of its host
at her feet, raises her eyes to the fleck
in the distance. We follow her.
We look wide, forensically,
deep into the mighty kelp forests
in the lower intertidal and subtidal strata offshore.

*Laminaria hyperborea*
*Laminaria digitata*
keystone species vital to birds
fish
epiphytes
crustaceans

the invertebrates they feed on
an entire ecosystem growing and replenishing
for eons

through the forest canopy we dive
past gulls foraging on kelp blades the length of orcas
down into the dusky under-canopy
we plunge
past the great northern diver
cormorant
eider
down past lobster and crayfish
butterfish and wrasse
down into the murk
the low glint of sticky eggs on stipes
nests of eggs in the chambers of the holdfast

hands first
we dig fingers into seabed
skin slippery in the suspension of kelp detritus
shed from blade tips
mucus
myriad spores
life-giving bacteria

feel your body lengthen
in this dark richness
limbs spread wide
your toes swell and splay
swaying in the shelter of the under-storey

flesh is thallus

# Glossary

Several of the seaweeds mentioned in Ellen Hutchins' letters have undergone name changes and/or reclassifications since Ellen's lifetime. In these cases the nomenclature given by Ellen has been used within this poem, and the current names are included below.

| | |
|---|---|
| bladder wrack | *Fucus vesiculosus* |
| dillisk (dulse) | *Palmaria palmata* |
| egg wrack | *Ascophyllum nodosum* |
| fork weed | *Furcellaria lumbricalis* |
| frond | "The leaf-like or erect part of the thallus" (Bunker *et al.*, p.27) |
| *Fucus amphibius* | Renamed to *Bostrychias corpoides* |
| *Fucus capillaris* | Renamed to *Gloisiphonia capillaris* |
| *Fucus cristatus* | Renamed to *Pterosiphonia complanata* (WoRMS) |
| holdfast | "Structure for attaching the thallus to the substratum" (Bunker *et al.*, p.27) |
| intertidal | "the intertidal is the area covered and uncovered by the tide with the upper shore immersed only 20% of the time, compared with the lower shore which is immersed 80% of the time" (Bunker *et al.*, p.10) |
| lamina | "blade of the seaweed" (BotRejectsInc) |
| *Laminaria digitata* | oarweed |
| *Laminaria hyperborea* | forest kelp |

| | |
|---|---|
| meristem | "area of the seaweed where growth occurs" (BotRejectsInc) |
| midrib | "conspicuous thick central line extending the length of the blade" (Bunker *et al.*, p.28) |
| stipe | "stalk-like portion of the thallus arising from the holdfast" (Bunker *et al.*, p.28) |
| thallus | "general term for the body of an alga" (Bunker *et al.* p.28) |
| thong weed | *Himanthalia elongata* |
| veined tongue weed | *Apoglossum ruscifolium* |

# Notes

I have chosen to keep the text of *Holdfast* free of citations or footnotes; nonetheless, as a work of research into a life lived two centuries ago, this poem owes much to a number of print and archival sources. Citations for all quoted and paraphrased material are provided below, listed by page number.

Chief among my sources are letters written by Ellen Hutchins herself. Ellen was a keen writer, relying on the exchange of letters and samples by post for collaboration with the fellow botanists of her time, and the materials cited below include extracts from her extensive correspondence with English botanist Dawson Turner. One letter by Matilda Hutchins, Ellen's sister-in-law, is also quoted. Many of the letters cited are published in Michael Mitchell's *Early observations on the flora of South West Ireland: selected letters of Ellen Hutchins and Dawson Turner, 1807–1814*, while one is published on the website of the Ellen Hutchins Festival. Some are unpublished; these were provided by Madeline Hutchins and are quoted by kind permission of the Ellen Hutchins Festival. For the sake of brevity, the writers and recipients of letters are referred to by their initials (EH, DT and MH) in the citations below.

# frond

2
*"I live an extremely retired life, next to complete seclusion & in such a situation it is rather doing me a kindness to excite exertion."*
Letter from EH to DT, 4 September 1809 (Mitchell, p.30)

4

*Yet the scalpel slid into this soft place, / opened the veins of mother and brother,*

"My Mothers age and the state of health of one of my brothers who has by a paralytic complaint lost the use of his limbs binds me so strongly to home that I am not likely to quit it." Letter from EH to DT, 8 October 1808 (Mitchell, p.20)

*"seed"*

"The fructifications resemble that of *F. hypoglossum* & when they burst the seeds are placed as in that species in lines each side of the midrib, I saw them burst." Letter from EH to DT, 15 December 1807 (Mitchell, p.13)

4–5

*They swap opinions on species, / on Dante and the newest Byron.*

"I am just going to begin Dante which I have not yet read" Letter (unpublished) from EH to DT, 4 July 1809

"I have just read *Childe Harold*. How do you like it? Parts of it delight me extremely. I am surer that Lord Byron can write fine poetry than that he is an amiable man. I don't like his severity towards woman & some other parts of his poem do not please my mind but are there not beautiful passages in it?" Letter from EH to DT, 7 December 1812 (Ellen Hutchins Festival)

5

*"at a time when domestic troubles / lean so heavily upon me."* Letter from EH to DT, 24 July 1810 (Mitchell, p.40)

*Her mind floats on pockets of enquiry: / the death of a friend and the drawing of* confervae *– / "new ones I hope" – / in the same paragraph.*

"The death of a friend has disturbed my mind a good deal

& affected my Mother so much as to make her worse than usual. She is in a very delicate state & I fear likely to remain so with perhaps some little daily variation of better & worse." Letter from EH to DT, 20 August 1811 (Mitchell, p.52)

*"new ones I hope"*

"Her being so has allowed me to spend parts of the last two days drawing *Confervae* (new ones I hope) for you." Letter from EH to DT, 20 August 1811 (Mitchell, p.53)

Atra *in rivers at Glengarriff*
Fucus capillaris
Fucus amphibius
Laceratus lingulatus (Mitchell, pp.83–84)

*notes the similarity of sea plant and furze / in terms of abundance and colour*

"Have you met with *F. caniculatus* [*Pelvetia canaliculata*] in an inflated state? It is so common a plant here that the shore is as yellow with it in summer as the land is with Furze." Letter from EH to DT, 20 August 1811 (Mitchell, p.54)

6

*pinky red purplish red / glossy brown reddish brown / dark blue green colour / olive brown beautiful green*

"It changes like most of the fine red plants to a purplish red." Letter from EH to DT, 15 December 1807 (Mitchell, p.13)

"The colour is very fine, pinky red, pale but bright." Letter (unpublished) from EH to DT, 14 March 1808

"The drawing of *Fucus fructilosus* will I hope meet your ideas of the nature of the beautiful original which abounds here. The color is olive brown which in drying changes to redish brown & in winter it is also redish." Letter from EH to DT, 30 January 1812 (Mitchell, p.55)

*"Its colour when recent is much more brilliant"*
"Its colour when recent is much more brilliant than after being dried." Letter from EH to DT, 15 December 1807 (Mitchell, p.13)

*colours change – / in spring / from age / after drying*
"*Conferva flocculosa* grows here in the sea on small *Fuci* & *Confervae*, it has when recent a rich brown colour, but changes in drying to green." Letter from EH to DT, 15 April 1808 (Mitchell, p.17)

*"2 pink coloured parasitic species, little beauties"*
"I wish you had the *Confervae* I have just drawn. I have 5 marine ones different from any I sent you, 3 of them veined & 2 pink coloured parasitic species, little beauties" Letter from EH to DT, 20 Aug 1811 (Mitchell, p.54)

# midrib

8
"*Your* rivularia vermiculata *(*mesoglioa vermiculata*) I am inclined to think not the same as my plant.*" Letter from EH to DT, 4 September 1809 (Mitchell, p.30)

9
*central "vein"*
"a large and beautiful veined variety." Letter from EH to DT, 8 October 1808 (Mitchell, p.21)

*Another, round and full when fresh, / flattens within minutes to show the inner thread.*

"When fresh it is perfectly round & has no appearance of midrib. After a few minutes passing it becomes quite flat & shows the thread that runs through it." Letter (unpublished) from EH to DT, 1 February 1808

*of which she sends a single plant / since she meets it rarely.*

"I have been so fortunate as to get a few plants of the ... *Fucus laceratus* that I sent you a single one of in a letter last year and of which I never saw more untill I met with a few this season." Letter from EH to DT, 8 October 1808 (Mitchell, p.20)

*In winter the leaf falls away, / leaving a single spike*

"In winter plants may have a very different appearance when the fragile blade lamina decays leaving only the midrib." (Bunker *et al.*, p.46)

11

*Ellen longs for / the early light of summer mornings – more time / to collect or write letters before breakfast.*

"How I long for summer mornings when I can have many undisturbed hours before breakfast." Letter from EH to DT, 10 January 1810 (Mitchell, p.33)

*Fine days, she takes a boat to rocks and strands, / returns, samples cosseted, hair thick with salt.*

"Mine was found on the 18 September and I suppose the fruit ripens in winter but as the rock where it grows is at some distance in a very exposed place and very difficult to land on except in the calmest weather I have not been able to go to it." Letter (unpublished) from EH to DT, 14 March 1808

*She makes "parcells", lays the fragile plant / on glass slides – easier to lift for the microscope.*
"... *conferva* & and another great beauty No 63 sent on glass in the last parcel." Letter from EH to DT, 10 January 1810 (Mitchell, p.32)

*Folklore tells of rock pool creatures, / half seaweed, half fish.*
"Bream: half fish, half seaweed" (Becker, p.139)

12
*bilious attacks*

"After being weakened by repeated bilious attacks succeeding each other so quickly that I had not time to recover strength I was seized with inflammation. When the fever left me I lay for many days & nights in so weak a state as to be only just sensible of existence. I can now get across the room without help ..." Letter from EH to DT, 11 October 1813 (Mitchell, p.64)

*"I am now so much better, tho' I have not been out for / more than two months untill this day."* Letter (unpublished) from EH to DT, 29 December 1810

## stipe

14
*"I have spent many happy hours creeping among its rocks & never quitted it without regret."* Letter from EH to DT, 10 January 1810 (Mitchell, p.33)

16
*"I never saw anything more beautiful than / its magnified appearance."* Letter (unpublished) from EH to DT, 4 February 1809

18
*Ellen speaks of her delight in this poetry, / and the horror of it.*
"Dante delights me horrible as he is tho' I only read the English translation." Letter (unpublished) from EH to DT, 4 September 1809

# holdfast

22
*"Send me a moss anything just to look at."* Letter from EH to DT, 30 November 1814 (Mitchell, p.68)

23
*never / turn your back on the sea.*
"Don't turn your back on the sea because you don't know how far she is going to come." (Becker, p.36)

*They used to make rafts, / roped the weed to float back to the village, / or rowed for an hour to get to the rock / where the strapweed is slender, / heavy with sap.* (Becker, pp.45-46)

*Ellen brings a hammer to take samples.*
"I think I have found *Urceolaria fimbrata* but have not gathered specimens as I had no hammer with me yesterday when I saw it for the first time here." Letter from EH to DT, 9 May 1811 (Mitchell, p.48)
\*Note this is a reference to the finding of a lichen, but

hammers were also used to collect seaweed samples if the holdfast was wanted.

25
*"My beloved Miss Hutchins breathed her last / in my arms the ninth of this month"* Letter from MH to DT, 26 February 1815 (Mitchell, p.68)

*Ellen's desire / that a parcel of plants be sent to him.*
"She desired that a particular parcel of plants should be sent to you. They were the best she had and very numerous." Letter from MH to DT, 26 February 1815 (Mitchell, p.68)

*"our sweet friend"* Letter from MH to DT, 26 February 1815 (Mitchell, p.68)

## epilogue

28
*"The Bay has been all summer full of Whales, / they came in after Herrings & sprats."* Letter (unpublished) from EH to DT, 9 November 1810

# Bibliography

Becker, H. (2000). *Seaweed memories: in the jaws of the sea*. Dublin: Wolfhound Press.

Bunker, A.R., Bunker, F.S.D., Brodie, J.A. and Maggs, C.A. (2017). *Seaweeds of Britain and Ireland*, 2nd ed. Plymouth: Wild Nature Press.

BotRejectsInc. (2014). *Seaweeds* [online]. Available from: https://cronodon.com/BioTech/Seaweeds.html [accessed 26 June 2023].

Ellen Hutchins Festival. (n.d.). *Ellen Hutchins – the story* [online]. Available from: https://ellenhutchins.com/ellen-hutchins [accessed 26 June 2023].

Ellen Hutchins Festival. (n.d.). *Ellen's letters* [online]. Available from: https://ellenhutchins.com/ellens-letters [accessed 2 April 2024]

Kelly, E. ed. (2005). *The role of kelp in the marine environment*. Irish Wildlife Manuals, No. 17. Dublin: National Parks and Wildlife Service, Department of Environment, Heritage and Local Government.

Mitchell, M.E. ed. (1999). *Early observations on the flora of Southwest Ireland: selected letters of Ellen Hutchins and Dawson Turner, 1807–1814*. Dublin: National Botanic Gardens, Glasnevin.

Guiry, M.D. (n.d.). *Pterosiphonia complanata (Clemente) Falkenberg* [online]. Available from: https://www.seaweed.

ie/descriptions/Pterosiphonia_complanata.php [accessed 5 November 2022].

WoRMS World Register of Marine Species. [n.d.]. *Fucus cristatus Withering, 1796* [online]. Available from: https://www.marinespecies.org/aphia.php?p=taxdetails&id=503731 [accessed 5 November 2022].

Unpublished letters:
Ellen Hutchins to Dawson Turner, 4 March 1808
Ellen Hutchins to Dawson Turner, 4 July 1809
Ellen Hutchins to Dawson Turner, 4 September 1809
Ellen Hutchins to Dawson Turner, 9 November 1810
Ellen Hutchins to Dawson Turner, 29 December 1810

# Acknowledgements

I have had so much support during the writing of this book. First and foremost, I must give my sincere thanks to Madeline Hutchins and the Ellen Hutchins Festival for providing me with access to archive material, and for including the launch of *Holdfast* in their festival programme for 2024. I would also like to thank Denis Murphy and the staff of Bantry Library for providing a venue for the launch. I would like to extend my heartfelt gratitude to Madeline Hutchins for her interest and for her feedback on the draft manuscript. I wish to thank the staff at the National Herbarium for allowing me access to many seaweed samples from around the period when Ellen Hutchins was active, and for their patience in answering my many questions regarding the process of collecting and preserving seaweed samples. I owe a debt of gratitude to Cassia Gaden Gilmartin and the *Channel* team for the immense care and professionalism with which they approached the publication of this book, and for their enthusiasm regarding the work throughout. Finally, I would like to give my sincere thanks to the Arts Council of Ireland for the Agility Award granted to me in order to allow me the time and space to write *Holdfast*.